COAL
in the Energy Crisis

By the Same Author

Auto Racing
Be a Winner in Baseball
Be a Winner in Basketball
Be a Winner in Football
Be a Winner in Horsemanship
Be a Winner in Ice Hockey
Be a Winner in Skiing
Be a Winner in Soccer
Be a Winner in Tennis
Be a Winner in Track and Field
Bicycling
Cleared for Takeoff,
Behind the Scenes at an Airport
Deep-Sea World, the Story of Oceanography
Drag Racing
Mopeding
Motorcycling
Passage to Space,
the Shuttle Transportation System
Pipeline Across Alaska
Project Apollo, Mission to the Moon
Project Mercury
Skyhooks, the Story of Helicopters
Skylab
Spacetrack, Watchdog of the Skies
Tankers, Giants of the Sea

COAL
IN THE ENERGY CRISIS

CHARLES COOMBS

William Morrow and Company/New York/1980

Permission for photographs is gratefully acknowledged:
Allis-Chalmers Corporation 85, 87, 106; Association of American Railroads, 10, 12, 77, 80; Bethlehem Steel Corporation, 40, 43, 47, 48, 50, 73 103; Carter Oil Company, 45, 51, 54, 62; Chicago Natural History Museum, 24; Coalcon Company, 114; Consolidation Coal Company, 44; Fluor Corporation, 116; National Coal Association, 8, 19, 32, 37, 101, 113; Southern California Edison Company, 88, 89, 109, 121; Tenneco, 91; United States Bureau of Mines, 28; United States Department of Energy, 16, 17, 18, 20, 39, 72, 79 92; United States Department of the Interior, 95; United States Steel, 36, 38, 41, 46, 82, 111. All other photographs by the author.

Library of Congress Cataloging in Publication Data

Coombs, Charles Ira, 1914-
    Coal in the energy crisis.
    Includes index.
    Summary: This discussion of coal highlights the changing technology of underground and surface mining, the controversial issue of land reclamation, and the importance of this fossil fuel's many by-products.
    1. Coal—Juvenile literature. [1. Coal. 2. Coal mines and mining] I. Title.
TP325.C578   333.8'22   80-13701
ISBN 0-688-22239-0   ISBN 0-688-32239-5 (lib. bdg.)

Contents

Early coal mining was dismal, dangerous drudgery.

1
The Forgotten Fuel

Coal has been known and used for many centuries. The Chinese dug coal about three thousand years ago. Greek philosophers mention its use around the Mediterranean area before the birth of Christianity. Around the ninth century the English were mining what they called "col" from shallow pits. Later the word was spelled *cole*, and about three hundred years ago it finally changed to *coal*.

But while many peoples knew of the dark mineral that burned, they made limited use of it. Prior to the eighteenth century, they relied instead upon plentiful wood for heat and upon wind and water to power their mills and small factories.

Then, during the late 1700s, the British developed the steam engine and hooked it up to various machines. In the face of dwindling supplies of wood,

In the past, railroad locomotives used large amounts of coal.

coal was called upon to provide the heat to generate the steam. Manufacturing increased in almost direct proportion to the amount of coal power available. The Coal Age had arrived, and the Industrial Revolution in Europe was under way.

Coal had long been known to exist in America, too, but again was little used here. Some Indians made black jewelry out of it. Some burned small amounts to fire their pottery kilms. Yet the most popular fuel was firewood, which was still abundant.

As steam engines were introduced into American industry, however, hotter and longer-burning fuels were needed. Coal answered the need, and miners went underground with picks and shovels to start wresting this valuable substance from the earth.

Two things happened in the mid-1800s to add impetus to the use of coal in America. One was the development of the railroad system that crisscrossed the United States. Giant fire-belching steam locomotives consumed great amounts of coal. In addition, the railroads moved mountains of coal from widely scattered mines to a growing number of customers. The railroads were so critical to the use and distribution of coal that for a time they virtually controlled the industry.

The second major factor benefiting the coal business was the conversion from wood to coal in the emerging steel industry. Carefully roasted coal, called "coke," proved enormously more efficient than wood charcoal in the making of steel. Soon steel plants were consuming about 15 percent of the special grades of bituminous coking coal being mined.

Coal remained the most important fossil fuel well into the present century. Not only did it run machines, it stoked the furnaces that heated millions of homes and factories. As the demand for electricity exceeded the supply available, primarily from hydro-

The railroad revolutionized the transportation of coal.

electric power facilities located along major rivers, many coal-fired plants were built. By the 1960s, most of the nation's electricity was being generated by coal.

Nevertheless, coal was notoriously dirty and difficult to dig, transport, and use. Mining was low-paying and dangerous work before proper wages and safety systems were introduced. Gradually the use of coal declined as Americans increasingly turned to oil and natural gas, which were cleaner, easier to handle, hotter burning, and more efficient. In a relatively short time, oil and natural gas replaced coal as king of the fossil fuels.

Cheap energy was the order of the day. People believed there were endless amounts of both oil and natural gas. American automobiles grew bigger and guzzled more gasoline, the prime product made from crude oil. Windowless factories and towering skyscrapers relied upon energy-consuming air conditioners and heaters to keep everyone comfortable. Hardly anyone thought to turn off a light in an empty room.

Conserving energy was considered unnecessary. Americans reasoned that even if American oil fields fell short of supplying the ever-growing demand, the Arab countries, Africa, South America, and other oil-and-gas-producing areas of the world would quickly and happily make up the difference.

In 1950, every man, woman, and child in the United States was using the amount of energy equivalent to that contained in nearly forty barrels of oil each year. At forty-two gallons to a barrel, that represented a massive amount of energy used per person.

By 1965, the figure had soared from forty to fifty barrels worth of oil annually.

Although prices increased somewhat in many countries, they remained modest in the United States, where about one-twentieth of the world's population was consuming one-third of the world's production of oil.

Then, in 1973, low rumblings of impending shortages erupted into a deafening roar. For political reasons, the Arab nations suddenly cut off their supply of oil to the United States. The embargo lasted through much of 1973 and 1974. The cost of crude oil quadrupled almost overnight.

Long lines developed at gasoline stations, and people complained bitterly over the sudden doubling of prices at the pumps. They scrambled for hard-to-get furnace oil. But these temporary inconveniences were soon forgotten when the embargo was lifted and imports of oil resumed.

By 1979, the annual per-capita consumption of energy in the United States had escalated to almost seventy barrels of oil. Once more there were warnings of shortages. Once more there were gas lines and increasing prices. For the first time in the history of the United States the cost of gasoline soared to more than a dollar a gallon. Although still cheap compared to prices in other parts of the world, the cost of filling large gas tanks became uncomfortably high in America.

The United States was importing half of the eighteen million or so barrels of oil consumed in the country each day. This proportion was disturbing, considering that any nation in the Organization of Petroleum Exporting Countries (OPEC) could cut off the flow at any time.

A simple answer to such dependence on foreign oil was conservation. If everyone used less energy, there would be enough oil and gas to go around. But too often conservation was expected only from the "other fellow." Another solution was to develop, improve, or expand on other sources of energy, but this course, too, brought problems.

Nuclear energy, whether fission or fusion, was encountering serious developmental and safety problems despite great promise for the future. These problems were highlighted in 1979 by dangerous radioactive leakage at the Three Mile Island nuclear power plant in Pennsylvania.

The exotic sources of energy, such as solar, geothermal, wind, wave, or tidal, required much time

Hydroelectric plants provide a small amount of the nation's electricity.

and money for development. Hydroelectric power, which accounts for only about 4 percent of the nation's energy, was difficult to expand also, since most rivers and streams were already dammed to capacity. Burning or fermenting the earth's grasses, grains, or other organic materials, known as biomass, or adding alcohol to gasoline to form gasohol showed some promise, but under current production techniques

Wind is an alternative energy source.

The Earth's inner heat can provide power.

could make only a slight dent in overall energy needs. Similarly, extracting synthetic fuels, called "synfuels," from oil-bearing shale and tar sands was enormously costly and difficult with established techniques. New development was necessary before either liquid or gas synfuels would be economically feasible.

So, out of all these alternatives, coal began to re-emerge as the best answer to the nation's immediate energy woes. In this critical period coal's reputation as a difficult and dirty fuel seemed less important.

Estimates vary as to the amount of coal the world contains. Since it is buried from a few to thousands of feet underground, any estimate can be no more than an educated guess. One calculation is that there are some eight trillion (8,000,000,000,000) tons of coal in the world. Approximately a third of that supply is located within the United States, in about three dozen of the fifty states.

United States Coal Deposits

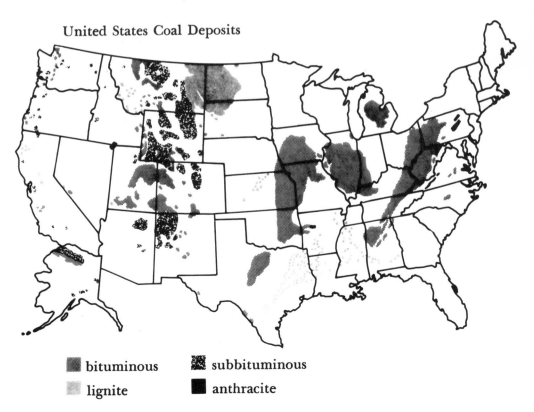

bituminous   subbituminous
lignite   anthracite

Modern mining techniques can reach some of the nation's coal.

But this figure is misleading, since most of the coal cannot be reached in any known manner or at reasonable cost. Thus, of the nearly three trillion tons estimated to be in the United States, only one-eighth, or about 350 billion tons, are recoverable. Although considerable amounts of oil and natural gas remain underground in the United States, coal makes up about 85 percent of the nation's known energy reserves.

With careful planning and controlled effort, this coal can be mined using present techniques. It is difficult to make an accurate count of the number of coal mines in the United States. Many are small, isolated, one- or two-man projects; others are tremendous holdings owned and operated by large companies. All told, they number in the thousands and are worked by well over 200,000 miners. The mines produce more than 700 million tons of coal a year. At this rate the supply will last about five hundred years. At a proposed increased production rate of a billion tons a year by the late 1980s, there still are some three hundred and fifty years' worth of coal reserves in the United States.

The figures are staggering, and they all point in one direction. Coal produces energy; coal's costs are in line with the climbing costs of oil and natural gas; there is plenty of coal to bridge whatever gap exists between the dwindling stocks of petroleum and gas and the potentially unlimited exotic energy sources of the future.

Coal, the forgotten fossil fuel, may be the energy source that can best meet present-day needs.

2

What Is Coal?

Coal is a product of the sun, the earth's primary source of energy. Some 300 million years ago, during a time called the Carboniferous period, the damp, tropical climate of a slowly cooling world gave birth to forests of giant grasses, broad-leafed vegetation, dense reeds, and other fast-growing plant life. These great plants and towering fern trees soaked up energy radiating from the sun. This energy was largely in the form of carbon, hydrogen, and oxygen atoms, which became locked within the leafy structures.

As the primeval forests matured and then died, the biomass, which also contained the trapped remains of reptiles and animals, accumulated layer by layer in the humid swamps. The decomposing vegetation formed into a dark, mudlike, woody mass, which later came to be known as peat. As the earth changed and the seas advanced and receded in cycles, these movements deposited layers of sand and clay atop the thick beds of peat. This material compressed the peat, squeezing out much of the water and sinking it deeper into the earth. In some areas, the earth's surface buckled and folded over the layers. The heat caused by enormous pressure drove out much of the oxygen and hydrogen, leaving a high concentration of carbon. This residue was coal.

During millions of years, the process repeated itself over and over. The amount of weight on the peat, and the heat generated by the pressure, largely determined the quality and characteristics of the coal. From five to more than ten feet of rotting vegetation eventually were compressed into one foot of coal. The greater the compression and heat, and the more hydrogen and oxygen driven out of the mass, the higher the remaining percentage of carbon. Coal is basically carbon. The more carbon, the harder the coal, and the higher its energy content.

Thus, over eons of time, large areas of the earth were streaked with veins, beds, or seams of coal of various thicknesses at various depths. The veins are usually separated by layers, or partings, of sand, clay, or rock. A seam might be an inch thick and a few

Coal was formed in ancient swamps like this re-creation.

acres in size while a bed could be more than a hundred feet thick and stretch for miles underground. Most coal deposits rank somewhere in between. They may lie hundreds, even thousands, of feet beneath the surface or a few feet below the tree roots.

Sometimes coal beds are level, marking the grassy swamps of the past. Other coal veins are slanted, tipped by ancient upheavals of the land, or upended by the folding of the earth's crust. There is no uniformity to the depth, size, or position of a coal seam.

Coal varies in quality as much as it does in location, depending on hardness, purity, and dryness. Theoretically, coal can be divided into four types: lignite, subbituminous, bituminous, and anthracite. Actually, one grade often can be substituted for another. However, the basic types do differ widely in the amount of water, sulfur, and other impurities that they contain.

The quality of coal is largely determined by the amount of heat energy that it will produce per pound. This feature is measured in British thermal units (BTU's). One BTU represents 252 calories, or enough heat to raise the temperature of one pound of water from 63 to 64 degrees Fahrenheit. Most soft coal contains from 70 to 85 percent carbon, the main heat-producing element.

Peat, which is not considered coal, is a soggy mass with little commercial value. Yet, when properly dried, peat gives off about two-thirds as much heat as low-quality coal. Peat bogs exist in many parts of the world. In the United States, peat is used as fertilizer. In rural parts of the British Isles and sections of Europe, peat is often dug up, dried, and used for stove and heating fuel.

Lignite is the lowest quality coal. Harder by far than peat, lignite has a high moisture content and gives off about 6,000 BTU's per pound. It is usually brown in color rather than the deep black of harder coals. Lignite has limited application as a boiler fuel in electric generating plants, which are the coal industry's biggest customers.

However, lignite can be liquefied or gasified into synfuels, which are gaining importance. Also, lignite serves as a petrochemical feedstock from which thousands of carbon- or coal-based drugs and products are made. Still, when better-quality coal is available, as it usually is in the United States, there is little incentive to mine lignite. Large deposits scattered from the Southern states to Canada are almost untouched, though they do provide a valuable energy reserve.

Subbituminous coal is the next higher grade. With about half the moisture of lignite, subbituminous is still relatively soft. But generally, like the best grades, it has a low content of sulfur, which is coal's most noxious pollutant.

Subbituminous, which is found mostly in Alaska

and Western United States, generates about 8,000 BTU's per pound. Vast seams lie near the surface of the Western prairies, where it is strip-mined and used primarily as a thermal coal to heat the boilers in electric generating plants.

Bituminous is the most plentiful kind of coal. Also classified as a soft coal, it is considerably harder than either lignite or subbituminous. Bituminous is found in great quantities both east and west of the Mississippi River, though the better-grade is in the eastern sector—Illinois, Indiana, Ohio, Pennsylvania, the Virginias, and south into Alabama. Much bituminous is obtained through surface mining, or strip-mining, techniques. Other vast deposits lie hundreds of feet underground and can be reached only by deep-mining methods.

Bituminous, containing between 10,000 and 12,000 BTU's per pound, fires the steam turbines in scores of electric generating plants from the Atlantic seaboard inland. Also, it can be converted into gas, light oils, and many carbon-based chemicals.

A large quantity of a particular grade of bituminous coal, known as metallurgical, or met, coal, is used to make coke, a major ingredient in the steel-making process.

peat

lignite

bituminous—bright coal

bituminous—splint coal

bituminous—
cannel coal

anthracite

coke

Different Kinds of Coal

A highly desired grade of coal, but the least common, is anthracite. This hard, very dense coal has the shiny, brittle blackness that is normally associated with the mineral. It can contain as much as 98 percent pure carbon. Anthracite burns the hottest of all coals, giving off about 13,000 BTU's per pound. Unfortunately, it is also the dirtiest. Anthracite has a higher sulfur content than some of the softer coals, and usually it requires special attention in order to meet environmental standards for clean air.

Anthracite was formed under the greatest geological pressure, and most of what is left lies in deep mines in eastern Pennsylvania. Once the most valued of coals, anthracite is now being replaced by the more plentiful and widespread bituminous due to its dwindling supplies, its geographic limitations, and the high cost of its transportation.

There are many shadings among the four major categories of coal, with lignite overlapping into subbituminous and bituminous being mistaken for softer strains of anthracite, but each serves some uses better than others. How wisely and efficiently these assorted grades of coal are used may determine how well the United States solves its energy problems.

3

Underground Mining

A skilled geologist can often tell just by the looks of the land whether a coal deposit is likely to lie somewhere beneath the surface. In hilly country particularly, guesswork sometimes is eliminated when outcroppings of coal poke out of the ground.

Most often, however, coal seams are discovered by drilling deep into the earth and taking core samples of everything that lies beneath the surface. A series of widely separated core drillings shows how deep the seam is located, how large it is, what the quality of the coal is, how much water may be in and around it, and whether the overburden, the ground above it, is stable enough to allow safe mining.

Several coal seams often lie atop each other, separated by layers of sand, clay, or shale. Thick and thin

30

seams alternate hundreds or thousands of feet into the earth. The total of all seams is called the "coal measure."

When the surveys are completed, plans are made for removal of the coal from the commercially feasible thicker beds.

There are two kinds of coal-mining operations, underground and surface, with numerous variations of each. Approximately half the coal dug in the United States comes from underground mines and half from surface, or strip mines, which are also called "open-pit" mines.

Underground mining, the more difficult and dangerous of the two, is used only when coal is buried too deep to be reached by surface-mining methods. There are three kinds of underground mines: drift, slope, and shaft.

Drift mining is the simplest of underground coal operations. When a seam of coal is discovered protruding from the side of a hill, a company sets up its mineworks on the spot. There is no time lost excavating through layers of rock. Instead, miners start digging straight into the exposed end of the horizontal coal seam. As their machines burrow deeper into the

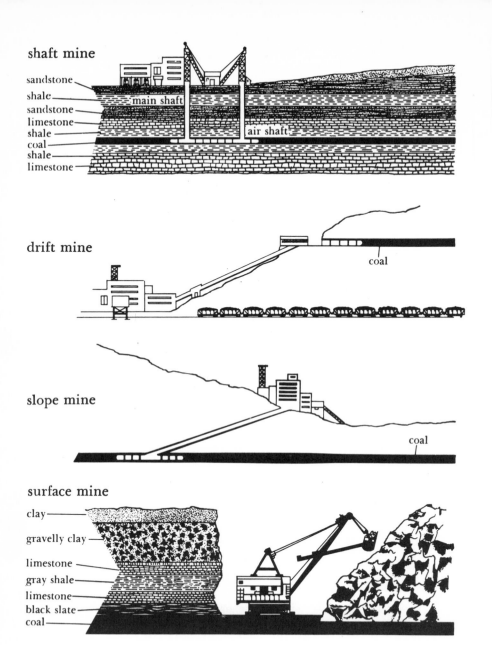

shaft mine

sandstone
shale
sandstone
limestone
shale
coal
shale
limestone

main shaft

air shaft

drift mine

coal

slope mine

coal

surface mine

clay
gravelly clay
limestone
gray shale
limestone
black slate
coal

Major Varieties of Coal Mines

hill, the coal is either transported out by low-slung, small railroad cars, called "tubs," or it is removed by endless conveyor belts.

Thus, as time passes, miners may find they have dug a mile or so into the mountain, following the drifting course of the coal vein. They will stay with it as long as the coal bed remains thick enough for them to operate their machines.

Often core samples reveal a coal seam that lies fairly near the surface, but too far down to be reached by strip-mining techniques. Since no exposed end is available for drift mining, a slope-mining operation is set up.

In a slope mine, digging machinery cuts a slanting tunnel downward through the overburden until it reaches the seam. Tracks are laid into the tunnel floor, and shuttle cars are run on them to transport miners and supplies to and from the coal face. Unless a conveyor belt is added, the coal also will be hoisted out in the low-slung cars.

As in any underground mine, giant coal-gnawing machines chew their way through the seam in all directions. Depending upon the extent of the seam, the mine may stay in production for years.

Of the three main types of underground mines,

the shaft mine is generally the most familiar. It is the classic coaling operation depicted in movies and books.

A shaft mine is usually begun by digging two widely separated holes straight down through the layers of sand, shale, or rock overburden to the coal seam. One shaft has an elevator, or cage, for transporting men and equipment into and out of the mine. Also, it is used to carry the forced fresh air to the miners working underground.

The second hole may be used as an exhaust shaft to remove stale air and dangerous gases, which tend to accumulate in coal mines. The all-important circulation system feeds fresh air down one shaft, fans it through the mine, and ejects it out another shaft.

Also, the second shaft may have a sump to collect the groundwater, always prevalent in deep mines, which at times constitutes a serious hazard to a mining operation and to the miners themselves. Powerful pumps working round the clock often are needed to keep a mine from flooding. Sometimes mines have to be abandoned when springs or underground streams are pierced and pumps are overwhelmed by rising water.

The second shaft, commonly called a "skip shaft," also is equipped with moving buckets, or skips, to haul the coal to the surface. There is no fixed pattern as to setup or use of the shafts. As long as proper ventilation is maintained, men, equipment, and coal can be moved quickly and safely.

A large proportion of shaft mines range around 300 feet in depth, but some are 1,500 feet deep or more. Whatever the depth, once a miner descends into the mine, he enters a damp, dark, and often dusty world.

Each miner begins his eight-hour shift by checking out his safety gear in a special equipment room. Often this room is a part of a so-called bathhouse where miners dress and shower at the start and at the end of their shifts. A few women work down in the mines, but by and large deep-shaft mining is dominated by men.

The miner dons his working clothes and hard hat with its firmly attached sparkproof electric head lamp. He wears safety goggles and gloves. He puts on reinforced boots to protect his feet from falling coal or other injury. He straps the cuffs of his coveralls so they won't catch in the machinery. Kneepads often are part of his equipment, for in low-ceiling mines he may have to work in a crouch or on his knees.

Each miner carries a safety lamp or other detector

A miner checks out safety equipment at the beginning of a shift.

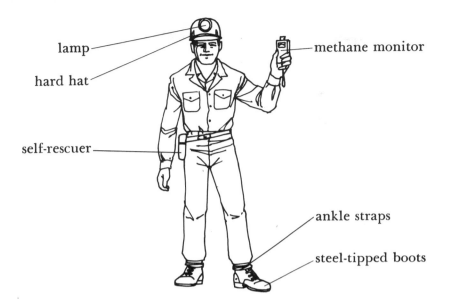

lamp

hard hat

methane monitor

self-rescuer

ankle straps

steel-tipped boots

to warn of poor ventilation or concentrations of dangerous, explosive methane gas. Each man also carries a respirator to filter out coal dust and a portable self-rescuer breathing apparatus to be used in an emergency.

Thus prepared, the miners enter the main-shaft cage and are lowered at high speed down to the coal seam. A large coal mine is much like an underground city, with "streets" and "avenues" providing access. It has an electric railway system complete with block signals and a dispatcher, radio and telephone communications, water and electrical services, and is fully air conditioned.

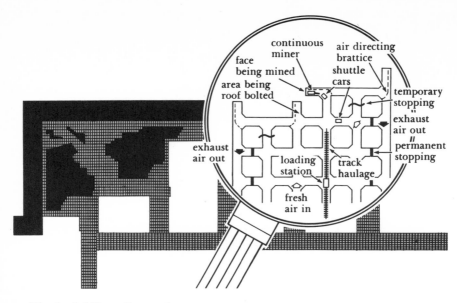

Typical Mine Operation

Upon reaching the bottom of the main shaft, the miners enter low, steel-topped, electric man-trip cars. These cars move along the tracks of the main street, or haulageway, which carries the mine's major traffic.

Near their working areas, the miners split up into crews and leave the man-trip cars. Walking, they disappear down various passageways, moving toward individual working faces. Most mines, depending upon their size, have several crews working on different coal faces simultaneously.

Oddly, the ceiling and walls around the worksite are white rather than black. As a precaution to settle the dust and prevent explosions, rock dusters coat

38

Underground miners climb into steel-topped man-trip cars, which carry them into the mine.

the tunnels with a limestone spray that turns the surfaces white.

Upon arrival at the working face, one miner immediately tests the air to be sure there is enough oxygen to breathe and no concentration of dangerous gas. Some other workers go about checking roof supports. Cave-ins are a miner's most feared and serious hazard, and the roof-control team is responsible for

Rescue teams stay prepared for emergencies.

preventing them. Some underground mines are still propped up by timbers, but most large, modern bituminous and anthracite mine roofs are strengthened by roof bolts.

As coal is removed and the roof is extended and

becomes less stable, roof bolters use ear-splitting pneumatic drills to bore holes several feet up into the overhead rock. Then they screw expansion bolts into the holes, to hold the rock tight and prevent cave-ins. Roof bolting does away with the more cumbersome and less effective use of props.

Roof bolting prevents cave-ins.

There are two main ways of removing coal from an underground mine: one is the room-and-pillar method; the other is the longwall method.

To understand the room-and-pillar method, one should again envision the mine as a city, with many blocks (the pillars of coal) separated by a grid of narrow streets. Machines move into the streets and start chewing rooms into the enormous blocks of black coal. The rooms expand as the pillars are eaten away. In time, even the many new rock bolts put into place will not safely support the ever-broadening ceiling. Accordingly, the machines stop removing coal while there are still pillars left to help support the roof. Although the pillars represent a considerable amount of coal that will never be brought to the surface, the safety they provide is essential.

Massive specialized machinery is used to remove the coal at the working face. In the so-called conventional method of mining, the coal seam is undercut by the long blade of a machine that resembles an oversized chain saw. A deep cut, called a "kerf," is made along the base of the seam. Vertical cuts and a kerf along the top may be added to block out a particular section of coal.

A member of the crew called a "shot firer" then

An explosives expert prepares to shatter a coal seam.

drills a series of holes into the seam and inserts explosive charges. The miners retreat to a safe distance while the explosive is set off. They stay out of the area until gases disperse and dust settles. Another more modern and safer method for dislodging the coal from the working face is to force compressed air through metal tubes to break it up.

In either case, the shattered coal is scooped up by

A loading machine scoops up shattered coal in a conventional mine.

a machine and loaded into small shuttle cars. These cars take the coal through the narrow passageways to the main haulageway, where it is either transferred to larger mine cars or, in the case of drift or slant mines, put on conveyor belts that carry it to the surface. There it goes through a preparation plant that grades, cleans, and sizes it according to the customer's order.

With the roof of the enlarging cavern properly bolted for safety, the coal machinery moves on to a new section of the working face.

Coal is cleaned, sized, and graded in a large preparation plant.

A continuous mining machine gnaws its way along a coal seam.

In room-and-pillar mining, the cut-drill-and-blast operation of the conventional system has largely been taken over by the continuous-mining method.

The continuous miner is a massive steel-toothed rotating head that gnaws away at the coal seam, moving loudly up and down and along the underground vein. Doing the work of a dozen men, the high-powered continuous miner can chew eight to twelve tons of coal a minute off the working face, scoop it up in gathering arms, and move it rearward into shuttle cars or onto a conveyor belt. Everything is done in one continuous operation, hence the name.

The continuous miner does away with the cutting

The rotating toothed head of a continuous miner is wet down to minimize dust.

Miner checking for gases

of a kerf and the drilling, blasting, and loading operations of conventional mining. However, the continuous miner creates a great deal of dust as it chews at
the coal, and large amounts of water must be spread
over the working face to keep the dust down.

Dust and methane-gas accumulations are constantly monitored during mining. If the readings

show more than two milligrams of dust per cubic meter of space, mining is suspended until the ventilation system cleans the air to the allowable standard. Special gauges on the continuous miner itself may cut off the machinery automatically if the accumulations reach a dangerous level.

Whereas the pick and shovel were the primary coal-mining tools of yesteryear, modern mining utilizes mechanized systems. But the room-and-pillar method still leaves approximately 20 percent of the coal untouched in the supporting columns. To get some of this coal, a more revolutionary method called "longwall mining," used widely in Europe, is being adopted in many deep underground mines in the United States.

In longwall mining, two parallel tunnels are bored horizontally to the far end of the coal seam, where they are connected by a cross tunnel. The mining begins at the cross tunnel, working backward to the starting place. The longwall miner is a tremendous machine made up of a powerfully driven, heavily toothed shearing wheel that moves noisily back and forth along the working face. It can plow a swath up to thirty-two inches deep during each pass along the face. Like the continuous miner, it drops the coal

A longwall mining machine moves back and forth across a coal face.

directly onto a conveyor belt, which transports it through one of the side tunnels to the mine shaft, where it is lifted to the surface.

The longwall miner differs greatly from the continuous miner in that special hydraulic ceiling supports move along with the machinery as it progresses through the coal, making a sort of walking roof. With this strong steel roof protecting men and machines from a cave-in, the water-bathed longwall miner cuts its wide swath through the coal. As the supports move with the operation, the ceiling of overburden left behind is allowed to cave in and fill the vacant space. The advantage of longwall mining is that there is no need to leave supporting pillars of coal standing; nearly all the coal can be removed.

Facilities above a large underground mine spread over many acres.

As in most other industries, automation has been introduced into underground coal mining, particularly in the case of continuous mining, which is the predominant method used in the United States. The machines are often remotely controlled by an operator watching a closed-circuit television, pushing buttons on an electronic console. The operator may be located in one of the underground roadways a safe distance away or even outside the mine.

An amazing array of special equipment and machinery is needed to wrest coal from the depths of the earth. A particular breed of man is also needed to work in the dismal, dangerous, and closed-in world of a coal mine. Often the product of generations of miners, these highly skilled underground workers produce nearly half of the coal used in the United States. As the demand increases, their roles in helping to cope with the energy crisis become increasingly important.

4
Surface Mining

Large areas of the United States, particularly in the West and Midwest, are underlaid with thick beds of coal. In some places in the far West as much as a ton of coal lies beneath a single square foot of surface area. This so-called steam coal makes excellent fuel for turning the giant turbines in electric utility plants.

Some of this coal is classified as bituminous, but most of the biggest deposits are low-sulfur subbituminous. Many large seams lie on top of one another, sandwiched between partings of clay, shale, or sandstone.

Sometimes only the main seam is mined. However, if the lower seams are not too deeply buried and are sufficiently thick and of good grade, mining continues layer by layer as long as it is practical.

53

upper seam

clay parting

lower seam

Different machines remove coal from a large, multilevel strip mine.

Most Western coal lies at a shallow enough depth to be removed by surface-mining techniques rather than by deep-shaft underground-mining methods. Fairly typical of a surface mine operation is that of the Black Thunder Mine near Gillette, Wyoming. Modern, as are many Western surface mines, the Black Thunder facilities are large, clean, and efficient.

A specially built, broad dirt road, packed to the

consistency of concrete, branches off the main high-way going south from Gillette and leads a dozen miles across the prairie. Browsing longhorn antelope lift their heads and stare at the passersby. Prairie dogs scold, and golden eagles wheel in the sky.

Beyond a gentle rise, the mine buildings, spread-ing over several dozen acres of ground, come into view. A railroad siding curves gently into the plant and passes directly under the biggest of several tower-

Covered conveyor belts connect the central load-out building to other preparation and storage facilities.

ing mine buildings. Some of the tall buildings are tied together by long, covered, tubelike structures that house large rubber conveyor belts. The highest belt stretches several hundred yards from the main tower, or load-out building, to a massive, roofed coal-storage facility. Other conveyor belts slant upward from the bottom of one machinery-filled building to the top of another.

In addition to the coal-processing structures, there is a modern administration building. There is also a large bathhouse for the convenience of the miners, where briefings and safety meetings are held. A large equipment repair shop, a security gatehouse, and a few other outbuildings complete the complex. Up to this point, there is no real evidence of coal.

Around a bend, however, hardly one half mile away from the offices, shops, and processing facilities, the surface-mining activity is well under way. Along a precipitous wall where the thin layer of rich topsoil and a thick layer of rocky overburden have been removed, a seam of black coal glistens in the sunlight.

The steps taken to locate, uncover, extract, and process coal from an open-pit, or strip, mine such as the Black Thunder are carefully planned and strictly controlled. First geologists and engineers prowl the area. They study the terrain, make land and aerial surveys, and analyze lengthy tapes of computerized data. They bore test holes and take core samplings. Satisfied that commercial amounts of marketable coal lie underground, the mining company signs land leases.

Then the company must make water-use agreements with state and local authorities. Water, which is essential for the mining operation, is precious in the prairie states and an adequate supply must be especially arranged for. Negotiations are held with the local utility company also to extend needed electrical service from the nearest power poles.

Next mining engineers determine the techniques that will best extract the coal and cause the least negative impact upon the ecology of the land. An environmental coordinator works closely with them. He tries to preserve water-runoff patterns, healthy air quality, welfare of the area's domestic livestock, and the habitats of such native wild animals as the golden eagle, deer, and antelope.

While potential problems are taken care of, the core drilling and analyzing continue. In time, a detailed map that shows the location, depth, and thickness of the coalfield emerges. The map measures the topsoil, plots the depth, density, and hardness of the rocky overburden, and shows variations in the quality, or grade, of coal at different locations. It spells out all that needs to be known in order to develop the mine.

Finally a starting point is chosen on one edge of the leased land or where the coal seam bulges conveniently close to the surface. Surveyors stake out the area where the first cut should be made.

In theory, the procedure is to dig down to the top of the seam, break up the hard coal with explosives, and then cart off the precious fuel. In reality, it is no longer that simple. Many new regulations and laws control modern mining. Despite the inconvenience and confusion these laws cause, their intention is to promote safe, sane, and clean mining practices. When done properly, coal can be removed from the earth without destruction of the surrounding land.

After all arrangements have been made, the bulldozers, scrapers, front-end loaders, draglines, power shovels, dump trucks, and other dirt-moving equipment arrive at the site where the first, or box, cut will be made. Some of the giant machines, such as draglines and power shovels, are assembled on the site as they are too big to be hauled on public highways.

A skilled worker operates a heavy dirt-moving machine.

Large machinery like this power shovel must be constructed on the mine site.

The first step is to skim off the topsoil, which may be a foot or more thick, and expose the top of the sterile rocky layer of overburden beneath. Trucks haul the topsoil to a special storage area, where it is reserved for future use.

Next come the blasters. They block out a sizeable area and drill a series of holes, approximately six inches in diameter, through the compacted overburden. Then they fill the holes with explosives, which either are contained in sausagelike sleeves or poured in from a bulk truck tank. A much safer mixture of ammonia nitrate and fuel oil is used instead of dynamite.

Though the giant power shovels can remove the overburden without first blasting it loose, the pre-

Valuable topsoil is stored separately and marked for later use.

Giant trucks cart away the overburden.

ferred procedure is to shatter the rock into small
chunks before scooping it up. Powder still is cheaper
than steel, and the mining company does not want
to wear out the big buckets. The shovel operators
also would rather work with less solid material.

With the explosives firmly in place and the crew
at a safe distance, sirens wail a warning. Someone
presses a button. A muffled roar sounds; the ground
heaves and belches dust. No sooner does the dust
settle than a big electric shovel moves in. With its

great steel-toothed bucket, it gnaws away at the loosened overburden and loads it into oversized trucks, which haul it off to a separate storage area. One special off-highway truck of monstrous size can haul up to 170 tons at a time.

The work continues day and night until the thick overburden is skimmed away and the top of the coal seam is exposed.

The process repeats itself. Blasters shatter the coal with more explosives. While one shovel works on the overburden, a second shovel digs into the coal and loads it into trucks, which rush it to the nearby processing facilities.

Preparing Western subbituminous coal for shipping or storage is less complicated than it is for the Eastern bituminous and anthracite varieties. The thick veins of subbituminous are relatively pure; the coal does not contain the clay that so often needs to be washed out of Eastern anthracite. And the Western coal is quite free of rocks.

The main processing that needs to be done is to reduce the random chunks of coal to a uniform size that is convenient to transport and will meet the customer's specifications.

At Black Thunder, trucks haul the coal directly to

Blasters plant explosive charges in the coal seam.

a primary crusher sunk below ground level. The trucks dump the raw coal—some pieces of which may be the size of a refrigerator—through a grating and into one of two 500-ton hoppers. Then they hurry off for another load.

From the hopper the coal drops onto an electric vibrating machine called a "grizzly." The grizzly screens out the smallest pieces, but feeds anything bigger than eight inches into a crusher. The crusher reduces these chunks to eight inches, about the size of a junior soccer ball. Although it is noisy and dusty, the crusher's location belowground helps muffle the sound, and mechanized dust collectors suck away the coal dust.

Reduced in size to a maximum of eight-inches in diameter, the coal now travels up a slanting, covered conveyor belt to a second crusher aboveground. En route a magnetic detector lifts out any tramp, or foreign, metal before it can get into the crusher. If the piece is too heavy for the magnet to handle, the machine automatically turns itself off. Tramp metal is not common, but a lost sledgehammer or a wheel lug from a giant truck could jam up the works and temporarily shut down the entire plant.

In the secondary crushing facility vibrating grills

Straddling an underground hopper, a giant coal truck prepares to dump its load.

again sift out the smaller chunks of coal and direct the eight-inch pieces into a finer crusher, which reduces them to a two-inch-diameter size. More dust collectors clean the air.

Now the coal, a mixture of sizes from two inches in diameter down to powder, travels up another long conveyor belt to the unit train load-out and sampling facility. The plant's tallest building, it straddles the railroad track.

When the coal, at a rate of up to 5,000 tons an

Coal dust, always present around a mine, surrounds a truck.

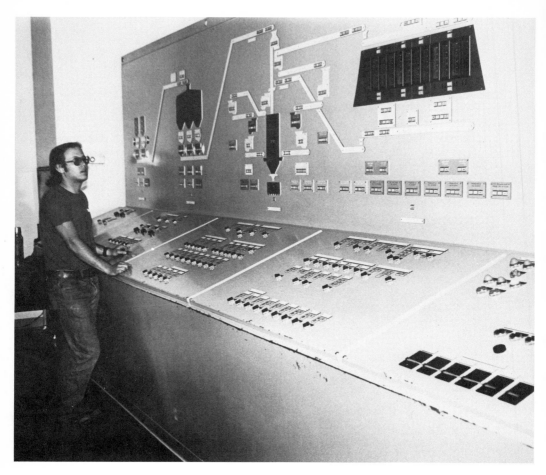

A computer directs coal through a plant or into trains.

hour, rides the conveyor belt up to the top of the
load-out tower, the load-out operator checks his
orders, scans the computerized control panel, and
decides where to route the moving coal. He may
trigger a flop gate and direct it straight down into the
slowly moving hopper cars of a waiting unit train. If

no train is available, the coal may be relayed by another belt to the big roofed-over storage facility, which can hold 100,000 tons of coal, enough to load up ten unit trains of a hundred large hopper cars each.

As the coal passes through the load-out facility, regular samplings are taken. Depending on quality, the coal may need special handling, selected storage, or perhaps require some blending of grades to give the buyer the type and quality of fuel he ordered.

During all operations, both around the mine and at the processing facilities, there is constant washing down and cleaning up.

Surface mining is less complicated than underground mining, although modern, mechanized methods of removal and strict environmental safeguards require much planning. Some mines use giant draglines to uncover the seam and tremendous self-propelled electric shovels to remove the coal. Some mines transport the coal directly out of the pit on conveyor belts. Almost every mine selects its machinery and varies its operation according to individual needs.

Some surface mining is done in mountainous terrain, where removing the thick, hilly overburden

Loading coal

Water trucks constantly dampen the dust.

can be impractical. The coal seam may show up as an outcropping on the side of the mountain, like icing inside a layer cake. Two popular methods, each a kind of contour mining, are used to remove this type of coal.

Where the overburden is not very thick, bulldozers may simply scrape it off and dump the waste material over the downside of the hill. The operation is much like excavating a roadway into the side of a hill. It

71

The dragline in the background removes overburden while the power shovel harvests coal from a lower level.

goes much deeper, however, at times removing most of the hill.

Once the coal is exposed, it is loaded onto trucks and hauled away. This method can be devastating to slopes, and particular care must be taken to reclaim the land as called for by the Federal Strip Mining Act.

Another method of contour mining, which is less

disturbing to the land but harvests less of the coal, is to use giant augers. Not unlike regular wood-boring bits in design, the coal augers may be three or four feet in diameter. Sections can be added, allowing the extended augers to reach some 200 feet into the vein of coal.

Augering is another surface mining technique.

Surface mining accounts for about half of the nation's coal production.

As the giant drill bores deeper into the seam, the coal emerges as large black shavings, which are loaded and carted away. This type of mining is limited to the depth the augers can reach into the hillside and to the relatively small amount of coal chips the augers are able to deliver to the surface.

Through a variety of methods, surface mining accounts for about half the coal produced in the United

States, approximately a half billion tons a year. In achieving this amount, the industry uses great ingenuity, plus some of the largest land-moving machines in the world.

Coal mining, whether underground or surface, is not a gentle business. Yet, in a time of critical energy shortage, coal may well be the alternative fuel needed to fill the gap between dwindling petroleum stocks and promising future power sources.

5

Moving Coal

Once the coal is mined, it must be transported to market. The United States digs out nearly a billion tons of coal a year, and more of it is shipped than any other single product.

Over two-thirds of this coal is moved from mine to customer in railroad cars. The distance may be a few miles or halfway across the country. Coal mines operate in three-fourths of the nation's states, and for purposes of economy a company tries to work out its distribution so that the coal travels as short a distance as possible to reach the customers.

Railroads carry coal in 100-ton open-topped gondolas or in large hopper cars with hinged bottoms for quick dumping. For customers who need only one or two carloads of coal at a time, the gondolas are put into trains made up of boxcars, flatcars, or

Vast amounts of coal are moved in unit trains.

tank cars and dropped off to buyers along the route. On a relatively small order, which requires special handling and attention, the price per ton can be high.

Big customers, such as steel mills or electric utility plants, purchase their coal by the trainload. These complete trains, called *unit* trains, may be made up of sixty to more than a hundred fully loaded cars, and they can carry more than 10,000 tons at a time.

A unit train shuttles between the mine and a single

customer. It travels directly between points, with no intermediate route changes or switching, and stops only for servicing or to change crews. At least one of them, an electric train, is completely automated and unmanned.

Typically, the attached cars of a unit train seldom come to a full stop. To load the coal, the cars move slowly under coal chutes at a preparation plant or storage yard. Creeping steadily along, an entire unit train can be loaded in less than an hour. Under careful traffic control, it moves onto the main line and picks up speed. Within hours or, at most, days, it is dumping its load, car by car, on a customer's stockpile. More often than not, that customer is an electric generating station or a steel mill.

Even during unloading, the train may never have to come to a full stop. While in transit, the hopper cars can dump the coal out their open bottoms onto a moving conveyor belt. This belt may then transport the coal on to crushers, which pulverize it, then feed it into the plant. More likely, the belt will move the coal to storage piles for later use in the steam boilers.

Unit trains made up of solid-bottomed gondola cars can be routed across rotating tipples, which turn each car over and dump its contents into bins. These

A unit train is loaded at a coal tipple.

bins feed giant conveyor belts. Gondolas equipped with swivel couplings need not even be detached from the train while the dumping takes place.

In order to handle the ever-increasing amount of coal being used, railroads are continually adding cars, strengthening roadbeds, laying new track, and updating traffic-control systems to prevent confusion and tie-ups.

Unit trains often are routed to harbors where the coal is transferred to sea-going ships and barges.

79

Most coal is transported by rail.

Many sea-going tows carry coal from ports in Texas and other Southern states across the Gulf of Mexico to coal-poor Florida.

Trainloads of coal also fill canal and river barges, which sail the nation's inland waterways. Many coal mines and preparation plants are located near major rivers. Barge transport is not only convenient, it is one of the least expensive ways of moving coal. Still,

80

although a barge can carry much more coal than a railroad car can haul, it is restricted in size because it must be able to squeeze through the locks on rivers and canals. For most of a trip, however, a dozen or more barges can be lashed together to form a string,

A dockside coal yard

Coal is also moved by barge.

or tow, which is pulled or pushed by towboats. When
locks are encountered, the tow is broken up, and the
barges pass through the lock one or more at a time.
Then the tow is reassembled and continues on.

In addition, lake and river freighters carry cargos
of coal, contributing to the total amount of several
million tons that moves over the nation's inland
waterways each year.

While railroads, freighters, and barge lines haul the bulk of coal throughout the country and to foreign ports, trucks, conveyor belts, and pipelines are widely used, too. Enormous off-road trucks frequently move coal from a surface-mining operation to an adjacent preparation plant where it is cleaned, sized, and graded. If trains are not handy, trucks also transport the coal to the nearest loading facility, dock, or storage area.

Many power plants are purposely built close to a mine, sometimes at the very mine mouth, in order to minimize the transportation distance. In these cases, trucks shuttling between mine and plant often do the carrying.

Fleets of trucks of a size permitted on public streets and highways deliver a fair quantity of coal to small users. The neighborhood coal truck disappeared from the nation's thoroughfares during the era when oil and gas virtually took over the heating of homes and factories. However, a few are coming back now, although they still are objects of curiosity in most areas.

Conveyor belts—long, broad, endless loops of rubberized material—have taken over a substantial portion of the task of moving coal. One conveyor belt

Covered conveyor belts control coal dust.

Conveyor belts distribute coal in a storage area.

in western Kentucky stretches ten miles from two mines to a barge loading dock along the Ohio River. This meandering, covered belt can transport more than 140,000 tons of coal a week.

In general, conveyor belts are much shorter, however. Most of them are used to lift coal from underground slant mines or out of drift mines and transport it to the preparation plant at the mine mouth. Many belts also carry coal from open-pit mines to railroad loading platforms or to nearby processing plants in place of trucks.

Quite commonly, where mine-mouth generators have been built, conveyor belts carry the coal directly to a conveniently located power plant. These endless belts are much in evidence wherever coal is mined or used. They greatly reduce the amount of necessary handling.

Pipelines also are being adapted or newly laid to handle coal. Transporting coal through pipes was first done in London nearly a hundred years ago. Slow to catch on but gaining attention, the process for piping coal has not changed. The coal is ground into a powder, then mixed with water to form a black soup, or slurry, which is pumped through pipes in much the same way as any other liquid.

Typical of slurry lines is the Black Mesa Pipeline in northern Arizona, which spans a distance of 273 miles. It begins at the coal preparation plant and reaches across desert and mountains to a 1,580,000-kilowatt power plant located in southern Nevada.

To prepare the slurry, the coal is first crushed into

Graded and sized coal travels by conveyor belt.

quarter-inch or smaller particles. A second step pulverizes the coal into a powder and mixes it with an equal amount of water.

The thick broth is then pumped through the pipeline at a speed of about four miles per hour. The controlled speed is important. To move it faster would cause excessive friction and abrasive wear on the in-

Powdered coal and water are mixed to form a slurry.

Coal slurry is stored in tanks before being dried.

side of the pipe. To move it more slowly would allow the coal to settle out of the solution, partially fill the pipe with a tarlike residue, and completely disrupt the flow.

Largely operated by Navajo and Hopi Indians, from whose lands the coal is purchased, several pumping stations propel the slurry through the eighteen-inch pipeline at the rate of about 4,200 gallons a minute, or approximately 660 tons of coal per hour.

When the slurry arrives at the generating plant after a three-day journey, it passes through whirling centrifuges, which spin out about 85 percent of the water. The recovered water goes into settling basins and is reused for cooling purposes in the plant.

The coal comes out of the centrifuge in a damp, cakelike form. It is then passed through a series of pulverizers, where grinding action and a continuous blast of heated air reduce it again to a dry powder. The same hot air that dries the coal blows it into the furnaces, producing steam at a temperature of a thousand degrees that drives the large turbine generators.

Although slurry pipelines are on the increase, they have their problems. The Black Mesa Pipeline is one that operates up to expectations, but many are disappointing.

Slurry pipelines are expensive installations, and only in certain instances can they be built and operated economically. First, there must be a constant need for coal at a single destination, such as a high-megawatt power plant. Second, there must be a sufficiently large supply of coal at the point of origin,

Transporting coal by pipeline is increasing.

the mine. Third, a large volume of water is required to mix the slurry. (In most Western and Midwestern areas sufficient quantities of water are hard to come by.) Fourth, the anticipated life-span of the power plant must be long enough to warrant the expense and effort of building a pipeline to it. If these criteria are not met, more flexible modes of high-volume transportation are usually better to use.

Converting coal to electricity enables it to be "sent by wire."

On the other hand, pipelines function continuously in one direction, thus eliminating the problem of nonproductive return trips. Also, they are an efficient way to transport coal after it has been converted to gas or liquefied into an oil product.

Finally, coal, or the energy from coal, can be transported by wire. In some areas, particularly in the

West, distances are great, and transportation facilities are inadequate. Instead of trying to haul the coal from the strip mine to some faraway power plant, the generating facility is built at or close to the mine mouth. The coal is processed on the site and fed directly into the power station's steam boilers. As the electricity is generated, it is sent silently and efficiently through high-voltage lines to distant markets.

Moving coal energy by wire adds no pollution to the urban areas where the electricity is used, requires no tracks or roadways, and eliminates the need to transport millions of tons of the bulky, dirty coal itself.

Over short or long distances, in single truckloads, unit trains, barges, ships, through pipelines, or by wire, coal is constantly on the move.

6
Healing the Scars

Mining coal has never been a clean job. In bygone days giant machines tore at the earth, leaving it ravaged. Coal-processing plants fouled the air. Contaminated water was dumped into the handiest drainage ditch or stream. When an area was mined out, men and machinery picked up and moved on to a new location.

But such careless mining practices are no longer possible. Mining operations are monitored by the Environmental Protection Agency (EPA). The Federal Strip Mining Act of 1977, the Clean Air Act, the Water Pollution Control Act, the National Environmental Policy Act, and a host of other regulatory measures leave little opportunity for slipshod procedures.

Costly and inconvenient as many environmental

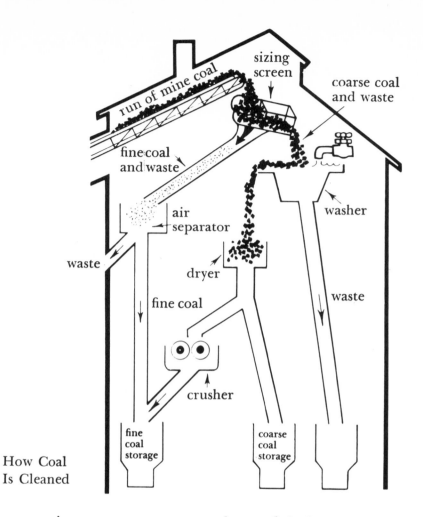

run of mine coal

sizing screen

coarse coal and waste

fine coal and waste

air separator

washer

waste

dryer

fine coal

waste

crusher

fine coal storage

coarse coal storage

How Coal Is Cleaned

protection measures are to the coal industry, every major American coal mine complies with them, although not always with a smile. There may be justification to some complaints that the coal industry is overregulated. In a time of critical energy shortage, mining the fuel and rushing it to market seems of prime importance. But, if improperly managed, the coal industry could cause lasting damage. Thus, great care must be taken in the mining, transportation, and use of coal.

Land reclamation is a major consideration wherever coal is mined. Even in the case of underground mines, considerable disturbance occurs outside the pit. Large amounts of clay and waste rock, called "gob," or "culm," collect from the mine and the preparation plant. These tailings used to be dumped on a pile and forgotten. The heaps of waste would erode in heavy rains or scatter in the wind, scarring slopes and polluting streams.

Today coal-mine wastes are carefully distributed, contoured, fertilized, watered, and replanted. After a few years former mine tailings are often difficult to identify. Sometimes the refuse is returned underground and used to refill the mined-out seam, restoring the environment to its original appearance.

Land reclamation is particularly important and difficult in the case of strip mines. Most high-production surface mines create great gashes in the earth. If left untended and unimproved, grazing land, farmland, wildlife refuge, parkland, and even living area would be destroyed.

As power shovels (left) remove the coal, trucks return the overburden to the mined-out area on the right.

Most strip mines begin reclamation and restora-
tion of the land shortly after the first ground is
broken, and this work keeps pace with the mining
operation. The keys to success are to disrupt the land
as little as possible and to restore it quickly. No time
should be allowed for the scars made by draglines,

97

Returning the overburden

bulldozers, trucks, or power shovels to become permanent.

As mining progresses, the loose overburden, or spoil, is collected and brought back by trucks, draglines, or conveyor belts to the hollow where the coal was recently removed. There it is dumped, and bulldozers and scrapers contour it to harmonize with the surrounding landscape.

Next the stored-up topsoil is hauled back and carefully spread out across the overburden. Care is taken to comply with rules against disturbing or polluting streams or upsetting long-standing drainage patterns.

To prevent erosion from rain or wind, native grass

Native grasses grow on restored strip-mined land.

and plant seeds are sown, fertilized, and watered as soon as the topsoil is down. Depending upon terrain and soil conditions, seedling pines, poplars, junipers, and other hardy trees or shrubs are planted. Given time, the results of reclamation efforts can improve grazing lands beyond their original state. When properly done, some land is sufficiently restored to be farmed.

Quite often, if the rainfall patterns and drainage systems are favorable or if an underground spring has been tapped, a new lake is established in a mined-out hollow. Surrounded by a newly planted forest, the restored area is suitable for camping, boating, fishing, or hunting. Some such places become wildlife preserves. Many new recreation spots have developed from carefully planned mine reclamation efforts.

Land reclamation is costly and time-consuming. Fortunately, it is now mandatory. Miners no longer can remove coal from the earth without a guarantee that full environmental safeguards will be taken.

Water often constitutes a problem around coal fields. There may be too much of it, particularly in the case of deep underground mines, where the coal seams lie below the natural water table and seepage causes flooding.

A herd of cattle graze on reclaimed strip-mined land.

Usually such excess water can be pumped out and dumped while the mining is taking place, but sometimes this water contains acid formed by the reaction of oxygen and iron sulfides in the coal and rock stratas. Such water drained into streams can kill fish. Spread over the ground, it can kill plant life. This acidity must be neutralized before the water is allowed into streams, spread over land, or otherwise

101

used. Thus, most mines have treatment plants to purify and reclaim mine water.

Whereas underground mines are often plagued with too much water, surface mines usually have too little. This condition is particularly true of the large open-pit mines located in the arid Midwestern and Western states. A lot of water is needed to wet down their operations and allay the dust, in order to decrease air pollution and keep working conditions tolerable. But the greatest amount of water is used in the coal-processing plants adjacent to many mines. When raw coal is not pure enough to be sized and sent to market readily, it is washed to remove the clay and other impurities. This operation consumes great volumes of water.

Where water is scarce, efforts are made to restore and recycle some of it. This job is often accomplished in settling ponds. Impure water is piped into large, earthen-diked reservoirs. After the solids settle out, the water is pumped back into use. With several such ponds to work with, one is settling while a second is filling, and a third may hold purified water.

In time, of course, the ponds fill with sludge. But they still need not be a loss. When the sludge has

A mixture of water, fertilizer, and grass seed is sprayed over graded mined land.

dried, it can be covered with topsoil and planted, so that it once more blends into the landscape.

Air quality is the last of the coal industry's primary environmental problems. Although there is concern about pollution created by all mining operations, the major polluters are the burners of the coal, the customers: electric generating plants, which consume about three-quarters of all the nation's mined coal, and steel, aluminum, cement, and other industries.

Dust, sulfur fumes, and the clouds of fly-ash particulates that pour out of smokestacks and spread in the wind cause the most concern. All coal contains varying amounts of sulfur, its most serious pollutant. When coal is burned, the sulfur combines with oxygen and forms sulfur oxides. These sulfur fumes are injurious to health, cause plant damage, deteriorate paint, and are generally hazardous. The emissions also contain excessive nitrogen compounds, which add to the pollution.

Western soft coal usually contains less sulfur than does the Eastern bituminous or the hot-burning anthracite. That difference is one reason why so many coal-powered electric generating plants are being built in the West. Considering the large amounts of coal burned, however, the cumulative emissions still are a potential hazard.

In a giant utility plant the coal is pulverized to a powder and blown with air into the furnace, where it burns continuously. The pollutants are ejected through flue stacks. If untreated, these impurities will spread throughout the atmosphere. Luckily, both the sulfur and fly-ash pollutants in coal can be reduced and neutralized to a large degree by an assortment of specially designed devices called "scrubbers."

Scrubbers are complicated and expensive additions to a coal-fired generating plant, sometimes accounting for more than 20 percent of the utility's overall construction costs. Most scrubbers treat the stack gases with limestone, which neutralizes the sulfur. Other types of absorbents remove the by-product, either discarding it as solid waste or treating it further to produce a salable material, such as sulfuric acid. Even the fumes can be turned into usable sulfur.

One method of scrubbing the pollutants out of burning coal is called a "fluidized bed combustion" (FBC) system. Limestone is mixed with coal as it burns. The sulfur and nitrogen pollutants combine with the lime, and then are removed in the solid bottom ash rather than going out the chimney and into the atmosphere.

Coal also can be made less noxious by cleaning

Maintaining good air quality depends largely upon clean burning of the coal.

it before it is burned. One expensive method is to separate the ground coal from the heavier sulfur-pyrite, settling them out according to their different specific gravities. A slightly less complicated system is to wash the coal in a frothy flotation bath that contains sulfur-removing chemicals. Still another means is to bombard the pulverized coal with microwave radiation and "cook" the sulfur out of it.

All of these processes are cumbersome and costly, but necessary for the protection of public health and the environment.

Perhaps the most promising means for reducing the pollution hazards of coal is to convert it into a cleaner fuel, such as a gas or a liquid, before using it. The impurities in the coal are removed and disposed of during the gasification or liquefaction process.

A great amount of money and effort go into developing methods for mining and using coal without endangering the environment. While the ideal would be zero pollution, the world cannot have all the energy it wants without paying something for it environmentally. All energy sources have their drawbacks. Even solar energy involves huge collectors, complicated converters, and high-voltage lines, which mar the landscape.

The apparent answer is to strike a reasonable balance between demand for economical energy and the preservation of a safe environment.

7
Coal at Work

In 1979, coal provided a scant 19 percent of the energy used in the United States; once it had accounted for 90 percent. Many mines had closed during the sixties and seventies, and out-of-work miners sought new occupations.

But in the late seventies the picture began to change. Most large energy consumers still preferred to burn cleaner, more efficient petroleum and natural gas, but these fuels were not always readily available and promised to become less so. As a result, coal mining began to increase; by the end of the 1970s, the annual production from both underground and surface mines in the United States exceeded 700 million tons and was on the rise.

Steelmakers continued to use vast amounts of coal in their blast furnaces as they need a ton of coal to

Most of the nation's electricity is generated in coal-fired plants.

produce a ton of steel. Today they have reached a level of use of about 18 percent of the total output.

Coal used in steelmaking is first converted into coke. Trainloads of special metallurgical coal are delivered daily to the nation's steel mills. Often it is blended with other grades of bituminous before being baked in batteries of narrow, airless ovens to temperatures as high as 2,000 degrees Fahrenheit. The intense heat drives off gas and tar, leaving a porous, gray-black, carbon-rich residue, which somewhat resembles barbeque charcoal. Shoved out of the ovens while white hot, the coke is immediately quenched with water to halt its burning. When dried, the coke is mixed with limestone and iron ore and burned in the blast furnaces to melt the iron out of the ore. Coke burns cleanly and with intense heat. It is as essential to steelmaking as the iron ore itself.

During the coke-making process, much combustive gas is released. Some of this gas is recirculated into the coking system and burned to add extra heat. Much of it, however, is purified and routed through pipes for commercial heating of homes, offices, and plants. The unwanted sulfur pollutant is salvaged and goes into the manufacture of insecticides, fungicides, acids, and other products.

The liquid tars distilled from the coking process are the most valuable of the coal by-products by far. Depending upon the distillation method used, the residue can be in the form of a heavy pitch, a light oil, or ammonia. In whatever form, the rich tars are processed into literally thousands of hydrocarbon-based items.

It can be safely said that no civilized person goes through a day without coming in contact with at

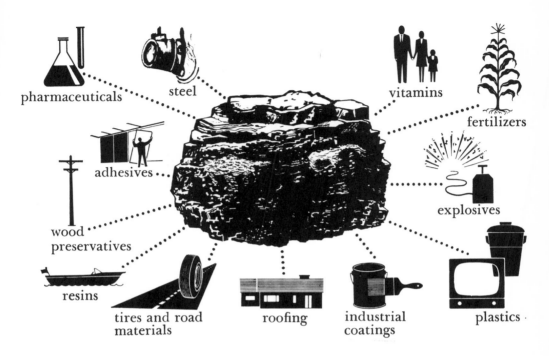

pharmaceuticals

steel

vitamins

fertilizers

adhesives

explosives

wood preservatives

resins

tires and road materials

roofing

industrial coatings

plastics

Products to Which Coal Contributes

least one product derived from coal. Coal is used to produce cement, paper, aluminum, roofing, or road paving. Many plastics are based on coal-tar chemicals. To walk on linoleum is virtually to walk on coal. A large variety of drugs and medicines are coal derivatives.

Food preservatives, dyes, detergents, nylon, synthetic rubber, nail polish, even mothballs are dependent upon coal. Coal plays a big role in pho-

111

tography, from making the film to developing and printing it.

Ammonia phosphates distilled out of the gases from the coking ovens enrich garden fertilizers. In fact, some of the phosphates go into the explosives that may have helped wrest the coal from the earth in the first place.

But the coal industry's biggest customer is the utilities, which generate the nation's electricity. Nearly three-quarters of all coal burned in the United States is used to generate more than half of the nation's electricity. Plants that switched to oil and gas when those fuels were cheap and plentiful are being converted back to coal. A ton of coal can save four barrels of imported oil, so converting about seventy-five oil-burning plants to coal-powered generators can save the United States more than 200,000 barrels of oil a day. Due to the plentiful supply of easy-to-reach, low-sulfur subbituminous coal in the West, many such coal-fired plants are being established there. Because of scrubbers and other cleaning equipment, these plants cost at least a third more to build than the simpler oil- or gas-fueled generators, but many consider them the best alternative today.

A giant coal-fired electric generating station

In its basic form, coal is bulky and difficult to handle. It is dirty and pollutes when it burns. However, coal need not be used in its native state. It can be converted to a gas, a liquid, or to another type of solid that burns cleaner and hotter.

Coal is mostly carbon and hydrogen with small amounts of oxygen, sulfur, and nitrogen. By adding more hydrogen, coal can be converted into a fuel

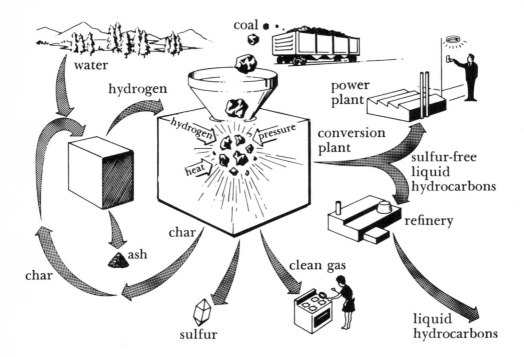

Through the addition of hydrogen, heat, and pressure, coal can be converted into gas or liquid fuels.

gas or liquid. Steam is added during the roasting process, and the steam, which is merely vaporized water (H_2O), supplies the hydrogen needed to make synthetic gas.

This substitute gas, often called "power gas," produces less heat than methane, or natural gas. But the coal gas burns cleanly, since its impurities, mostly sulfur, are removed during its conversion.

Unfortunately, synthetic gas loses still more of its heat-producing power when it is piped any distance. Such gas is more effective to burn close to where it is generated. Accordingly, coal gasification facilities and the electricity-producing turbines that use the gas are best located near each other.

Coal gas can be strengthened by mixing it with natural gas to increase its methane content. Or the gas can be put through an additional process called "methanation" in which still more hydrogen is added to increase its potency.

There are several different coal gasification processes being explored. All use heat and assorted amounts of hydrogen and oxygen. Since the latter two elements are obtained from water, a gasification plant is apt to use twice as much water as coal during its processing. This requirement creates a problem in the West, where strip-mined coal is plentiful but water is scarce. Still, as manufacturing techniques improve, gasified coal will be used more and more to heat homes and factories and to make steam. Easy to handle and clean burning, such hydrogen-fortified synthetic gas is highly desirable.

Liquefaction, or converting a solid into a fluid, is another way to transform coal into a cleaner, more

A complicated plant built close by a South African coal mine liquefies coal into various petroleum products.

usable fuel. Liquefying coal is not new. For a time during World War II, Adolph Hitler virtually ran his Nazi war machine on methanol derived from liquefying coal. Oil-poor South Africa has been producing thousands of barrels of gasoline a day from coal for years. Kerosene, or coal oil, has been produced in large quantity for more than a century. In recent years, however, kerosene, the foundation of jet fuel, has been distilled from petroleum rather than coal. Now kerosene is emerging from coal-liquefying plants again as are other fluid fuels.

In somewhat the same manner as gasifying coal, liquefaction requires the addition of considerable amounts of hydrogen to the carbon. Normally coal has a ratio of about sixteen parts of carbon to one part of hydrogen. In order to liquefy it into a hot-burning fuel, the ratio needs to be lowered by about two-thirds. For instance, fuel oil, a prime product of coal liquefaction, has a six to one ratio of carbon to hydrogen.

There are several complicated chemical processes for adding the hydrogen to coal so that a volatile liquid fuel is created. Although the cost of producing these synthetic oils is much greater than the cost of natural fuels, the lack of adequate petroleum supplies leaves little choice.

Both gasification and liquefaction of coal leave a great deal of solid waste, some of which can be used. Char, for instance, a carbon residue that emerges from at least one liquefaction process, can be burned economically after most of its sulfur has been removed. Mixed with a binder and pressed into briquettes, char becomes barbeque charcoal. However, much of the residue, or slag, from synthetic-fuel plants has little salvage value and must be disposed of. Some of it serves as nonpolluting landfill, or it is dumped back into the mine as spoil.

In another process, called "solvent refining," coal is transformed from one solid to another. Again, under heat and high pressure, crushed coal is mixed with a hydrogen-rich solvent such as a light oil. The coal dissolves. The ash and most of the sulfur is filtered out. The fuel that emerges is like a solid tar. It can be pulverized much like normal coal, or it can be melted and transported through pipelines. This so-called solvent refined coal (SRC) burns nearly twice as hot as raw coal, and it burns cleanly.

The major disadvantage of synthetic gas or oil is cost. Produced by complicated processes in expensive plants, synfuels are not competitive with other forms of energy today. Still, much effort is being made to bring these costs down. A major advantage of synthetic liquids or gases is that users can retain their plants without having to make costly conversions to burning solid coal.

Much coal exists in seams that are too thin or under overburden that is too thick to be mined economically. Billions of tons of this inaccessible coal lies deep in the earth. One way to reap the energy from these sources is to burn the coal where it lies.

Though the concept may seem like science fiction, it is a reality. The process is called *"in situ"* which means "in place," or right where it is *situ*ated.

The basic idea is to bore two holes down into the coal formation. One hole is used to supply the air that contains the oxygen essential to combustion. The second hole removes the product of the slow combustion—gas. In order to complete the circulating system, a third hole is drilled through the coal formation horizontally, connecting the other two. Oxygen is forced down into the underground deposit, and the coal is set to smoldering. Under controlled burning, large amounts of flammable gas are given off, piped, and put to use.

An in situ plant also must be equipped to clean the synthetic gas as it emerges from the ground. This work is done by scrubbers that use water or chemical mixture to remove the sulfur and nitrogen pollutants and to filter out most of the solid particles. The purified gas can also be treated with extra hydrogen to boost its methane content and increase its heat-generating efficiency. Or it can be put through

119

a pipeline and blended with natural gas to make it more suitable for use by consumers.

In all instances, the primary reason for converting coal to a gas, liquid, or solid is to provide the customer with an efficient heat-producing product. The synfuels are easier to transport, simpler to use, and generally superior to raw coal. By and large, synfuels can be readily substituted in plants that have been burning oil or natural gas without having to make costly alterations.

All avenues that will lead us to the full use of coal are being explored. In a time of critical energy need, some sacrifices may have to be made. Energy, by its very nature, is a volatile product, and a degree of risk and inconvenience exists wherever it is used in large amounts. But care must be taken to maintain reasonable controls over any operations dealing in the production and consumption of energy. While the relaxation of some environmental restrictions may be called for, potential hazards must be contained. Energy production and environmental protection should go hand in hand.

In one form or another, coal occupies an increasingly important place in the energy plan. It will be much in evidence during coming years, at least until

Coal will continue to fuel most of the nation's electric plants within the foreseeable future.

better, safer, and cleaner sources of energy are fully developed.

Although coal may no longer be considered the king of fuels, it still plays a royal role.

Glossary

anthracite: hard coal.

biomass: organic material, vegetable or animal.

bituminous: soft coal.

British thermal unit (BTU): a measure of heat, 252 calories.

char: a residue like charcoal.

clean coal: low-sulfur coal.

coal measure: sum of all seams in a deposit.

coke: porous product that results from roasting coal in oxygen-free oven.

continuous mining: a cutting machine rips coal from the face and automatically loads it into cars or conveyors.

conventional mining: older method of blocking out, breaking up, and loading coal while propping up roof.

conveyor belt: endless belt for moving coal or overburden.

culm: gravelly mine waste.

drift mine: mine that opens into a level or nearly level seam of coal.

face: solid surface of coal being worked on.

fossil: ancient remains of plants and animals.

gasohol: fuel mixture of gasoline and alcohol.

gob: hard mine-waste material.

grade: rank, size, and special treatment of coal.

haulageway: underground mine roadway.

headgear: works and pulleys at top of shaft to operate elevators.

in situ: processed underground, where it lies.

kilowatt (kw): 1000 watts.

longwall mining: system for mining a long, straight coal face.

lignite: soft, brownish coal.

man-trip cars: low cars for transporting miners underground.

megawatt (mw): 1000 kilowatts; one million watts.

methane: coal gas.

mine-mouth plant: utility situated near the mine.

open-pit mine: see strip mine.

outcrop: where coal appears on the surface of the ground.

overburden: the material that overlies a coal deposit.

parting: layer of earth between separate coal seams.

peat: wettest of the combustible decomposed plant fuels.

pillars: blocks of coal left in mine to support the roof of overburden.

preparation plant: where coal from the mine is cleaned, sized,
and mixed.

rank: grade of coal.

raw coal: coal that has not been cleaned or sized.

reclamation: renewing strip-mined land.

roof bolts: long expansion bolts that prevent roof cave-ins.

room: an underground area from which coal has been mined.

seam: a large deposit or layer of coal.

shaft mine: underground mine reached by a vertical shaft.

shearer: machine that cuts coal from the working face.

shuttle car: electric vehicle that transports coal from the working face.

slope mine: hillside mine reached by slanting tunnel.

strip-mining: mining across the surface of the land.
subbituminous: next coal grade softer than bituminous.
surface mining: see strip mining.
synfuel: synthetic fuel made from various nonpetroleum products
tip: mine waste dump.
underground mining: mine reached by shaft or tunnel.
unit train: long train carrying coal to a single customer.
Watt (W): unit of electric power, 1/746 horsepower.

Index

indicates illustration

About the Author

Charles (Chick) Coombs graduated from the University of California, at Los Angeles, and decided at once to make writing his career. While working at a variety of jobs, he labored at his typewriter early in the morning and late at night. An athlete at school and college, Mr. Coombs began by writing sports fiction. He soon broadened his interests, writing adventure and mystery stories, and factual articles as well. When he had sold over a hundred stories, he decided to try one year of full-time writing, chiefly for young people, and the results justified the decision.

Eventually he turned to writing books. To date he has published more than sixty books, both fiction and nonfiction, covering a wide range of subjects, from aviation and space, to oceanography, drag racing, motorcycling, and many others. He is also author of the Be a Winner series of books explaining how various sports are played and how to succeed in them.

Mr. Coombs and his wife, Eleanor, live in Westlake Village, near Los Angeles.